Collectible
AUNT JEMIMA

Handbook & Value Guide

Jean Williams Turner

Schiffer Publishing Ltd

77 Lower Valley Road, Atglen, PA 19310

DEDICATION

To Lynn and Kevin Burkett of Hillsdale, Michigan for sharing their wonderful collection and vast knowledge; to Myla Perkins for pictures and insisting I pursue this project; and to my husband John for sharing my interest in the life of Aunt Jemima.

ACKNOWLEDGMENTS

A special thanks to Charles Warren for photographing the Burkett Collection.

Title page photos: Aunt Jemima full face logos through the years. (From left to right) 1893-1917, 1917-1932, 1933-1968, 1969-1988, and 1988-present.

Copyright © 1994 by Jean Williams Turner
Library of Congress Catalog Number: 94-65621
All rights reserved. No part of this work may be reproduced or used in any forms or by any means – graphic, electronic or mechanical, including photocopying or information storage and retrieval systems – without written permission from the copyright holder.

Printed in Hong Kong.
ISBN: 0-88740-644-0
We are interested in hearing from authors with book ideas on related topics.

Published by Schiffer Publishing Ltd.
77 Lower Valley Road
Atglen, PA 19310
Please write for a free catalog.
This book may be purchased from the publisher.
Please include $2.95 postage.
Try your bookstore first.

AUNT JEMIMA PANCAKE FLOUR

"I'se in town, Honey!"

1919 magazine advertment, early logo, "I'se in town, Honey!"

Early 1900s shipping carton, with full face logo. 13-1/2" x 14". *Author's Collection.* $100-$150.

CONTENTS

From the December, 1896 issue of *Ladies Homes Journal,* an advertisement for Aunt Jemima's Self-Rising Flour. *Burkett Collection.* $10-15.

CHAPTER 1
The Aunt Jemima Story
"The Most Famous Colored Woman In The World"

Aunt Jemima's story began in St. Joseph, Missouri in 1888, when Chris Rutt, an editorial writer for the St. Joseph *Gazette,* and Charles Underwood, a friend in the milling business, purchased an existing mill along the Black Snake Creek and formed the Pearl Milling Company. The two partners developed a self-rising pancake flour. The name "Aunt Jemima" was the result of Rutt's hearing a cakewalk tune by the same name performed at a Baker & Farrell minstrel show by a character costumed in an apron and bandanna in the tradition of southern cooks.

And so the life of Aunt Jemima began. Rutt and Underwood organized their company and the Aunt Jemima trademark was registered. However, their success was shortlived. By 1890, without merchandising success and no capital, the company faltered.

Within a few months the partners reorganized into the Aunt Jemima Manufacturing Company, with Underwood's brother Bert as financier and promotor. Bert achieved no more success than his brother and Rutt had. Rutt returned to the *Gazette* and Charles Underwood found employ-

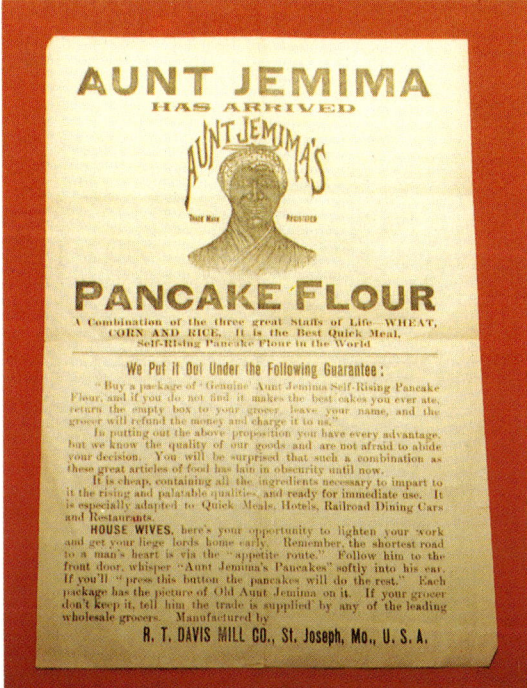

"AUNT JEMIMA HAS ARRIVED"— A very early handbill printed in black only, introducing Aunt Jemima and the newly created Pancake Flour manufactured by R. T. Davis Mill Co., St. Joseph, MO USA, ca. 1893. *Burkett Collection.*

"HOUSE WIVES, here's your opportunity to lighten your work and get your liege lords home early. Remember, the shortest road to a man's heart is via the 'appetite route'. Follow him to the front door, whisper "AUNT JEMIMA'S PAN-CAKES" softly in his ear."

Lapel pin, marked "Duryea & Co./Sole Agents/West Broadway, New York" on the image, "Whitehead & Hoag/Newark, NJ July 17, 1894" on the pinback. 7/8". *Author's Collection.*

ment with R.T. Davis Milling Company, the largest miller in St. Joseph. Shortly thereafter, Davis purchased the recipe and set out to improve and promote the product.

With the addition of rice flour and corn sugar to improve the flavor and the addition of powdered milk for simplicity, the mix was ready for promotion.

Davis envisioned a living Aunt Jemima to demonstrate the self-rising pancake flour, and to bring the trademark to life. Bring it to life they did — with friendly, 59 year old Nancy Green, born as a slave on a plantation in Montgomery County, Kentucky. She *became* Aunt Jemima. For more than 30 years, until her death on September 24, 1923 as the result of being struck by a car on East 46th Street on Chicago's South Side, she presided at promotional events for Davis across the country, including the 1893 Chicago World's Fair. Accompanying Davis to the Fair was Purd Wright, advertising manager for Aunt Jemima. He had designed a souvenir lapel pin with the likeness of Aunt Jemima and the caption "I'se in Town Honey."

Following the Fair, Wright created Aunt Jemima's folklore story, some factual, in a souvenir booklet entitled "The Life of Aunt Jemima, The Most Famous Colored Woman in the World." The booklet tells that Aunt Jemima was born "near the junction of the Red River with the Mississippi, on the left bank of the Father of Waters," in a plantation cabin. At an early age she was noted as a cook, unsurpassed in the preparation of certain dishes. Her Mistress, Mrs. Higbee, prized her as a household jewel in the Governor's Mansion, as Col. Higbee's dwelling was known before the war. Aunt Jemima's pancakes, with a combination of wheat, corn and rice cereals, were celebrated in the neighborhood. At the Governor's Mansion she cooked for many famous people of this continent and Europe.

As the story goes, Aunt Jemima befriended Confederate troops during the war. It was not until after the war that her fame as a cook accidentally reached the world, when an ex-Confederate general en route to New Orleans aboard the *Robert E. Lee* said to some acquaintances, "the best meal I ever ate in my life was at a Negro cabin not far from where we are now." When the steamer stopped for wood, the curious party of seven, led by the ex-general, ventured out for Aunt Jemima's cabin. It took her only a minute to prepare a batch of her famous pancakes. In the party was a representative of R. T. Davis Mills Co., who made a mental note of the location of the cabin, and on reaching New Orleans notified the firm of his discovery. He was instructed to "secure the recipe."

"It is said that nothing created so much of a stir among the Negroes of Louisiana in that year, 1886, as the sale of Aunt Jemima's Pancake Flour recipe to the representative of the R. T. Davis Mills Co...," the booklet explains. In her deal with the mill, Aunt Jemima made a couple of stipulations: 1) she be paid in gold, as she did not understand why U.S. bank notes were better than Confederate money, and 2) she be taken into the employ "so as to superintend the mixing of the ingredients that make up the pancake flour."

The booklet tells Aunt Jemima's story even after the sale of her recipe. At the 1893 Chicago World's Fair it was said "no exhibit of a food product created so much of a stir as that of R. T. Davis Mills Company." Their exhibit consisted of a giant flour barrel 12 feet across at the end, 24 feet long, and 16 feet in diameter in the center. Inside were an office and reception parlor for visitors. Around the parlor were different "medals and diplomas taken by the flour in all portions of the world at various exhibitions." But the biggest attraction of all was Aunt Jemima herself a few feet away from the barrel, making pancakes from Aunt Jemima Flour, "each package of which bears her portrait."

"The consequence was that at times the crowd was so great around this exhibit that the assistance of special police had to be secured to keep it moving," and "over 500,000 orders alone were received at the booth for

Original paper doll family, ca. 1895. A family of paper dolls were reproduced in the late 1970s by Ralph Griffith of Parkland, MO. *Photograph courtesy of Myla Perkins.*

packages of Aunt Jemima's Pancake Flour." Orders came from Europe, Canada and all over the United States, and R. T. Davis carried off first premium medal for "absolute purity and superiority" for Aunt Jemima's Pancake Flour.

In conclusion, the booklet explains that Aunt Jemima's Pancake Flour's phenomenal success is due to the guarantee on every package:

> "Buy a package of Genuine Aunt Jemima's Self-Rising Pancake Flour, and if you do not find it makes the best pancakes you ever ate, return the empty box to your grocer, leave your name, and the grocer will refund the money and charge it to us."

In 1895 new packaging was printed with a cut out paper doll of Aunt Jemima's family consisting of Aunt Jemima, Rastus, Abraham Lincoln, Delsie, Zeb and Little Dinah — showing them "Before the Recipe was Sold," barefoot and in ragged clothing, and "After the Recipe was Sold," finely attired.

An advertising campaign from 1919, penned by James Webb Young of the J. Walter Thompson Co., became known as the Legend of Aunt Jemima. These folklore tales were illustrated by N. C. Wyeth and were reproduced in four colors, appearing for the next 10 years.

A decade later an Aunt Jemima rag doll family emerged, and through the next years several versions appeared featuring Aunt Jemima, with Uncle Mose replacing Rastus and only two children, twins Diana and Wade.

In 1900 R. T. Davis died and several disasters were endured by the company. Following the bankruptcy of one of Davis' heirs, a new man was chosen to head the reorganization — Robert Clark, former general manager of R. T. Davis Mills Company. Clark put the business back in order, and in 1914 it was renamed The Aunt Jemima Mills Company.

Aunt Jemima Mills again reorganized after the Commodity Market Collapse of 1920 and it struggled for several years with the personal tensions of the management families. On January 15, 1926 Aunt Jemima Mills was sold to The Quaker Oats Company for a reported $4,202,077.28. Robert Clark continued with the business until his retirement in 1937. Quaker quickly expanded the Aunt Jemima product to retail stores all over the country, and the product experienced healthy growth.

During the depression of the 1930s Quaker's advertising people brought Aunt Jemima back to life to revive the public's interest, as Davis had succeeded in doing at the Chicago World's Fair. Want ads went into Chicago papers inviting auditions. It was Anna Robinson, a massive, 350-pound woman "with the face of an angel" (Marquette, p. 153) who portrayed Aunt Jemima from the Chicago Century Progress Exposition in 1933 to the early days of television, until her death in 1951.

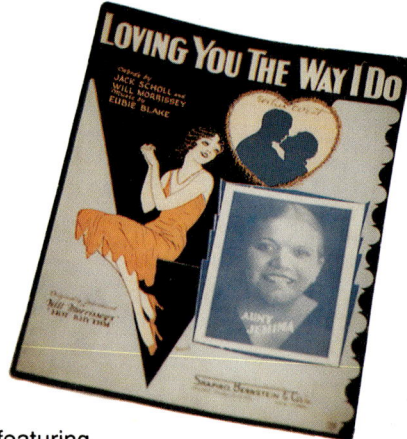

Various pieces of sheet music featuring or including Aunt Jemima with such noted performers as Shirley Temple, ca. 1920s.

Commercial artist Haddon Sundblom was commissioned to paint a portrait of Anna Robinson. That likeness appeared on the newly rede-signed Aunt Jemima packaging until the 1968 makeover.

Through the mid-twentieth century several women portrayed Aunt Jemima at promotional events. Among them, Amanda Randolph, a regular on the "Amos and Andy Show" as King Fish's mother-in-law. She also appeared in "Shuffle Along" in the 1920s with Josephine Baker.

In 1955 at Disneyland, California, the Aunt Jemima Kitchen was opened, where diners could eat pancakes and relive the days of Tom Sawyer. Here Aylene Lewis portrayed Aunt Jemima as a gracious hostess, "far removed from the days of servitude" of her counterpart Nancy Green, "lending dignity to the image of the Negro as America's unrivaled culinary expert" (Marquette, p. 139). Until her death in 1964, Aylene Lewis greeted tens of thousands of visitors from all over the world.

Since the Aunt Jemima trademark change in the 1930s, we have seen two other makeovers. In 1968 we saw the bandanna replaced by a headband. Then, in July of 1989, the headband was traded for soft, curly, gray-streaked hair, pearl earrings and a new lace collar.

The Aunt Jemima story spans more than a century — a century of faces, starting with Nancy Green, and a century of ease of preparation for housewives, from the addition of sweet powdered milk in the late 1800s to the introduction of today's frozen waffles, ready to eat after just a minute in the toaster.

Aunt Jemima's face and famous pancakes are as much of an American institution as apple pie!

A *Good Housekeeping* ad explaining the creation of 'Ready-Mix', November, 1916.

A 1916 magazine advertisement for Aunt Jemima Pancake Flour, Aunt Jemima Mills Company, featuring Aunt Jemima's Rag Doll Family.

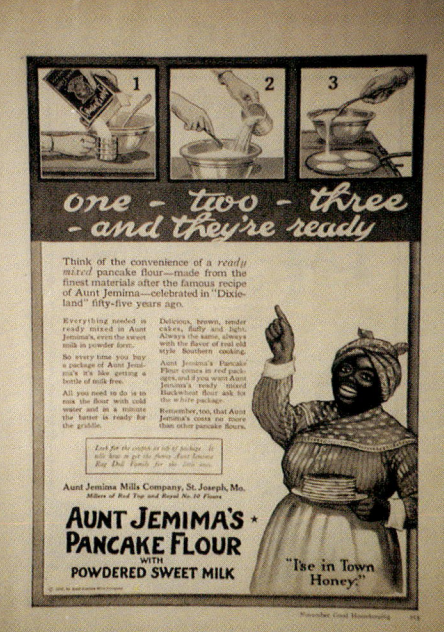

CHAPTER 2
Earliest Promotions

"Life of Aunt Jemima—The Most Famous Colored Woman in the World" by Purd Wright, ca. 1895. *Author's Collection.* $150-200.

"Aunt Jemima's Special Cake and Pastry Flour Recipe Booklet," compliments of Davis Milling Company. Booklet is die-cut like an actual box and both front and back covers are alike, ca. 1906. 3-1/4" x 4-7/8". *Author's Collection.* $150-$200.

Ladies Home Journal advertisement for Aunt Jemima's Pancake Flour. Information for obtaining Aunt Jemima and Her Rag Doll Family, *Special Cake and Pastry Flour Recipe Booklet,* and Climbing Aunt Jemima Doll, October 1, 1910. $25-$50.

String Climbing Doll—"Le' me to it! It's worth the climbing for Aunt Jemima's Pancake Flour." Marked "Germany." Piece is two parts cardboard which bend in the middle to climb up the string, ca. 1910. Doll 13", pancake box 3-3/4" x 2-1/4". *Burkett Collection.* $1200-$1500.

11

Above and Below:
Aunt Jemima's Pancake Flour Halloween Mask with Rag Doll Family advertisement on reverse, ca. 1902-1916. Another version of the mask also exists. *Author's Collection.* $450-$500.

Front Covers

Two versions of the Aunt Jemima Needlebook "A Household Necessity," with information on how to get the Cloth Rag Doll Family, ca. 1905. *Burkett and Author's Collections.* $125-$150.

Back Covers

Opposite page , Right and Below: "Fiddle and Bow" Ready Mix Self Rising Flour, Aunt Jemima Mills, St. Joseph, MO. Front, inside and back covers of needlebook, ca. 1914-1925. *Burkett Collection.* $100-$125.

Made from FIDDLE AND BOW FLOUR

BISCUIT PROBLEM SOLVED.

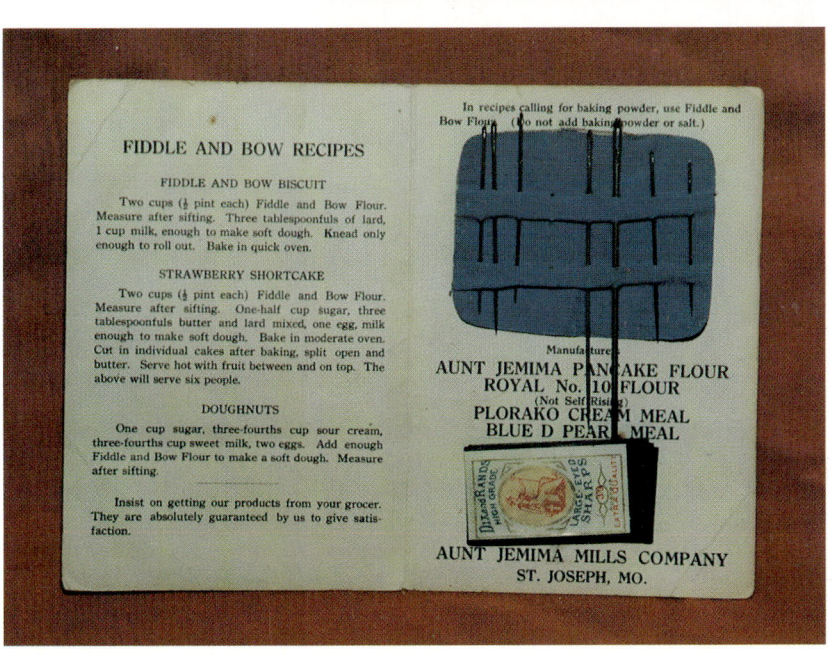

In recipes calling for baking powder, use Fiddle and Bow Flour. (Do not add baking powder or salt.)

FIDDLE AND BOW RECIPES

FIDDLE AND BOW BISCUIT

Two cups (½ pint each) Fiddle and Bow Flour. Measure after sifting. Three tablespoonfuls of lard, 1 cup milk, enough to make soft dough. Knead only enough to make soft dough. Bake in quick oven.

STRAWBERRY SHORTCAKE

Two cups (½ pint each) Fiddle and Bow Flour. Measure after sifting. One-half cup sugar, three tablespoonfuls butter and lard mixed, one egg, milk enough to make soft dough. Bake in moderate oven. Cut in individual cakes after baking, split open and butter. Serve hot with fruit between and on top. The above will serve six people.

DOUGHNUTS

One cup sugar, three-fourths cup sour cream, three-fourths cup sweet milk, two eggs. Add enough Fiddle and Bow Flour to make a soft dough. Measure after sifting.

Insist on getting our products from your grocer. They are absolutely guaranteed by us to give satisfaction.

Manufacturers

AUNT JEMIMA PANCAKE FLOUR
ROYAL No. 10 FLOUR
(Not Self Rising)
PLORAKO CREAM MEAL
BLUE D PEARL MEAL

AUNT JEMIMA MILLS COMPANY
ST. JOSEPH, MO.

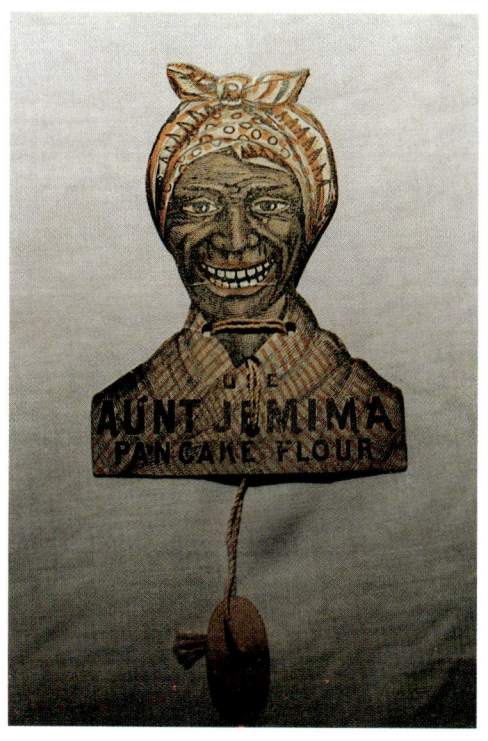

Above and Below:
Two-piece Aunt Jemima String Puzzles are found in three versions with advertising on reverse, ca. 1905-1916. *Burkett and Author's Collections.* $125-$150 if complete.

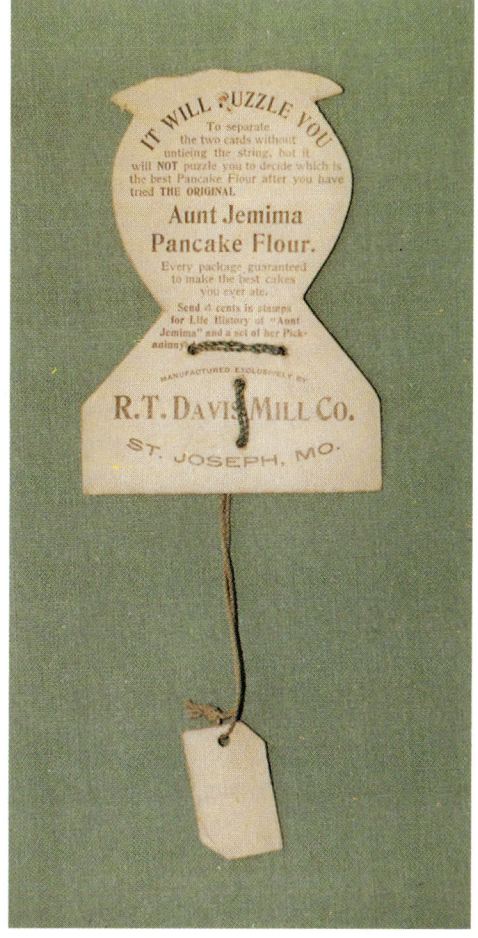

Above and Right:
"Send 4 cents in stamps for 'Life History of Aunt Jemima'," the cover photo booklet, ca. 1905-1916. $125-$150.

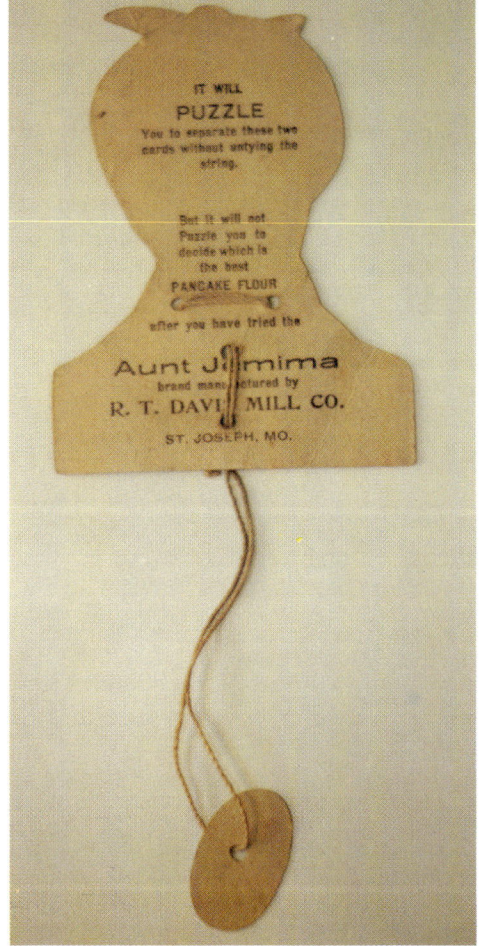

Above and Right:
Third version of string puzzle ca.
1905-1916. $125-$150.

"THE AUNT JEMIMA SLIDE," published in St. Joseph, home of Aunt Jemima Pancake Flour. Pieces of this sheet music are found printed with "Compliments of R. T. Davis Mills," suggesting that it was used as a promotional item, ca. 1917. 11" x 14". *Burkett Collection*. $50-$75.

"JEMIMA'S WEDDING DAY," another sheet music piece showing the use of the Aunt Jemima name and image, ca. 1899. 11" x 14". *Burkett Collection*. $50-$75.

CHAPTER 3
Packaging Through The Years

All photographs in this section are from
the Burkett Collection.

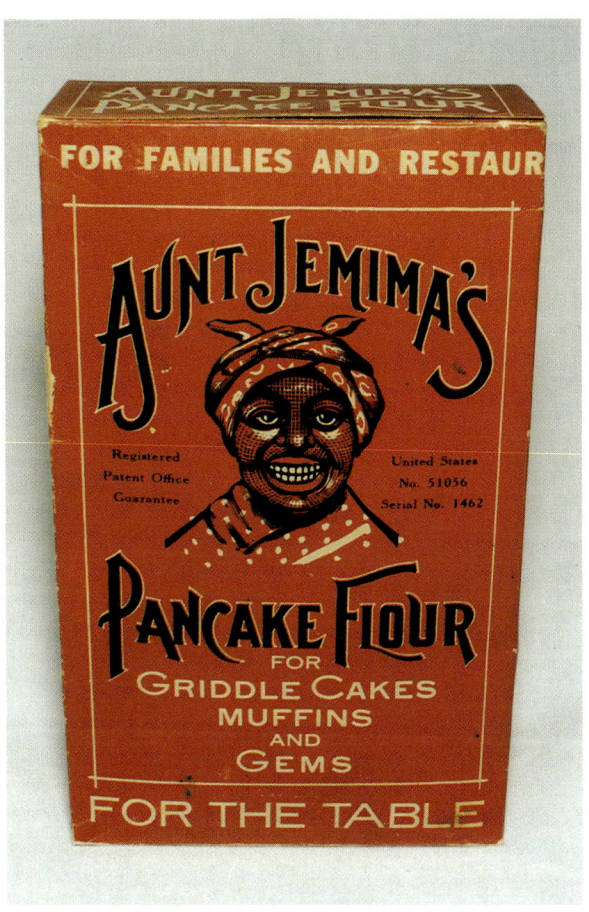

The following product containers show
the changes in the Aunt Jemima image
over the decades—starting with the
image on the wooden shipping crate
depicting the early version of Aunt
Jemima by R. T. Davis Mill and Mfg. Co.
to the cornmeal container with a friendly,
smiling Aunt Jemima by Quaker Oats.

Aunt Jemima Pancake Flour wooden crate, R. T. Davis Mill & Mfg. Co., St. Joseph, MO. Early full face logo on both sides, ca. 1890-1914. $200-$250.

"Cooking & Salad Oil" tin, ca. 1920s. Five gallons, 13-3/4" x 9". $500-$750.

A sample size box of Aunt Jemima Buckwheat, Corn & Wheat Flour. 4 oz. $125-$150.

Cardboard shipping cartons from Aunt Jemima Mills Co. and The Quaker Oats Company, after 1926. $100-$150 each.

Aunt Jemima Pancake Flour and Aunt Jemima Buckwheat, Corn & Wheat Flour Pancake Flour mix boxes, after 1926. Both boxes are 1 lb. 4 oz. $150-$175.

Aunt Jemima Pure Wheat Bran
bag, after 1926. 24" x 39".
$100-$150.

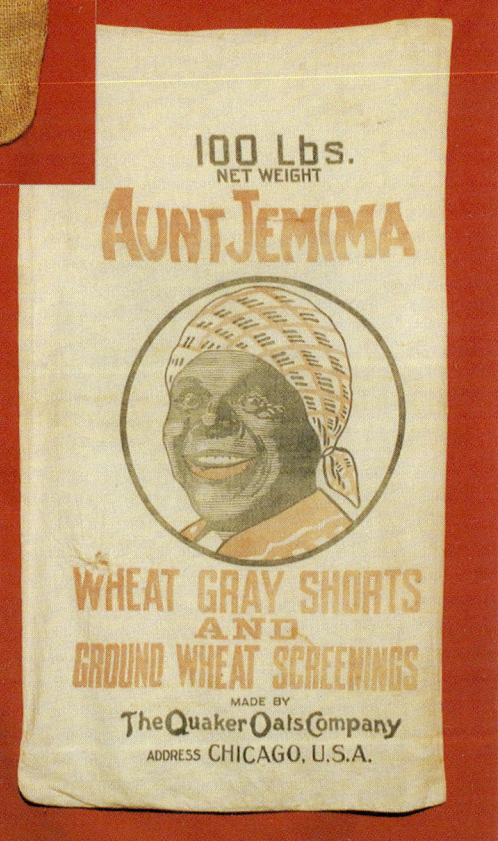

Aunt Jemima Wheat gray shorts and
Ground Wheat screenings bag, after
1926. 21" x 38". $100-$150.

Aunt Jemima Cornmeal bag,
after 1926. 16" x 10". $100-
$150.

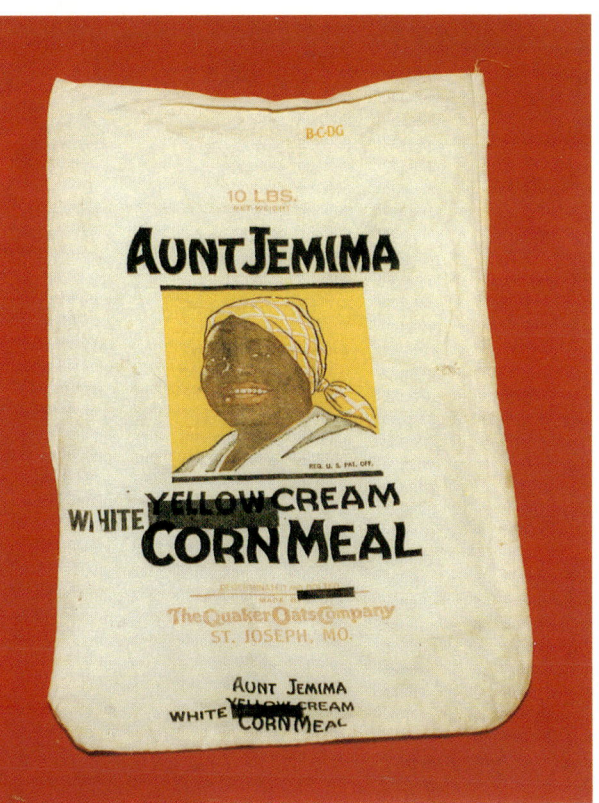

Aunt Jemima Wheat gray
shorts and Ground Wheat
screenings, after 1926. 24" x
39". $100-$150.

delicious for supper too!

REG. U.S. PAT. OFF.

AUNT JEMIMA
READY-MIX
for PANCAKES

WHEAT, CORN, RYE AND RICE FLOUR
PREPARED WITH PHOSPHATE, CORN SUGAR, SODA, SALT AND
POWDERED SKIM MILK
BY THE QUAKER OATS COMPANY, ADDRESS — CHICAGO, U.S.A.

Opposite page, Top and Right:
Three product boxes from the
late 1940s-early 1950s. $100-150
each.

All containers on pages 28 and 29 were made prior to the 1968 logo makeover. Pricing is dependent on condition.

Top Row: Pre 1968 logo.
Bottom Row: Post 1968 logo.

Aunt Jemima Pancake Mix in Spanish,
packaged for sale in Mexico, mid-1980s.

CHAPTER 4
Rag Doll Families

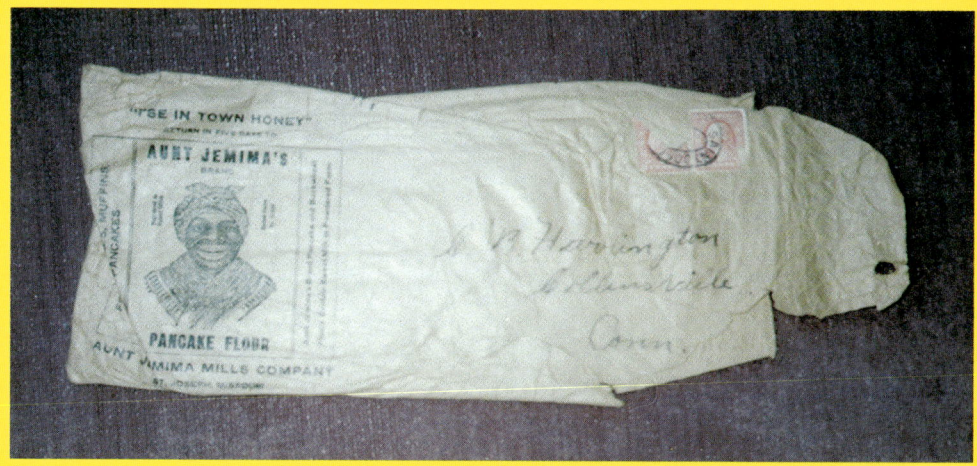

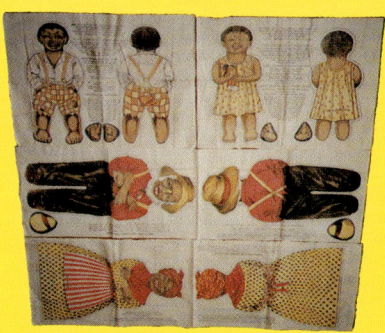

Aunt Jemima Rag Doll Family in uncut form with original mailer *1993 Christmas gift to the Burkett Collection.*

Dolls in uncut form with identifying background fabric bring much higher prices in today's market. In unsoiled condition, like these pieces, a collector can expect to pay $175-200 per doll. Dolls are often found cut and unsewn, sewn and un-stuffed, as well as stuffed. Today's prices range from $150-$175, depending on condition.

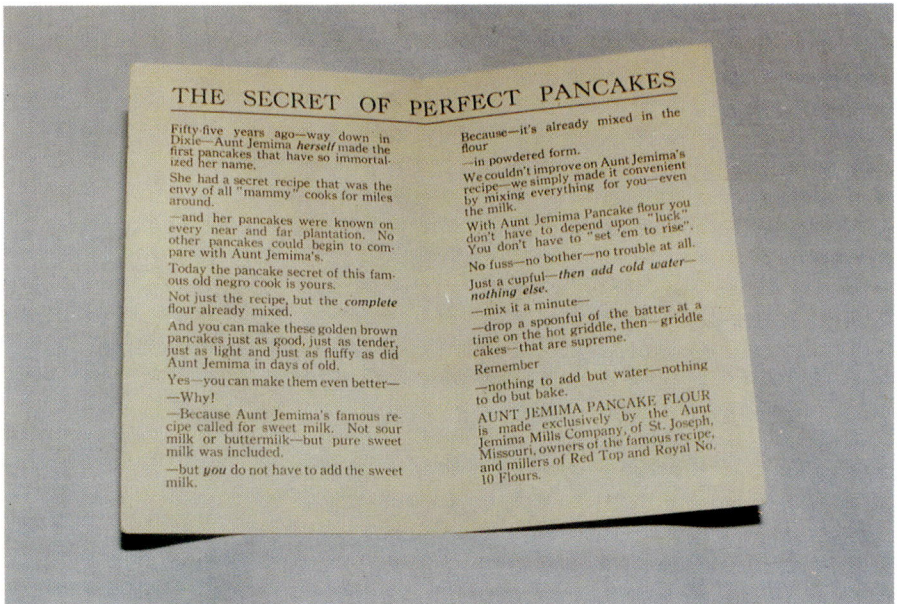

"THE SECRET OF PERFECT PANCAKES," Aunt Jemima Mills. Back of booklet introduces Buckwheat Cakes and instructions for ordering the rag doll family, copyright 1916. *Burkett Collection.* $75-$100.

A rag doll family coupon with a brightly colored family on the front, ca. 1917. 3-1/2" x 2-1/4". *Burkett Collection.* $25-$50.

Opposite page:
Davis Milling Co. magazine advertisement for Aunt Jemima's Pancake Flour, featuring a rag doll family offer for 3 boxtops and 10 cents in coins or stamps, 1916. *Author's Collection.* $20-$25.

AUNT JEMIMA'S
PANCAKE FLOUR

"Made in a minute,
The Milk's mixed in it."

AUNT JEMIMA'S PANCAKE FLOUR is ready mixed with pure powdered milk, giving that rich flavor to griddle cakes, muffins or waffles made with this delicious blended flour.

AUNT JEMIMA
RAG DOLL FAMILY

There's a coupon on top of every package of Aunt Jemima's Pancake Flour or Buckwheat Flour telling just how to get the funny Rag Doll Family. Every child wants these dolls, they are so droll and so good to play with.

Aunt Jemima

Magazine advertisement with a special offer for a "Jolly Family of Aunt Jemima Rag Dolls," December, 1924. *Author's Collection.* $20-$25.

PLACE
STAMP
HERE

WE HOPE THAT YOU ENJOY THIS BOOK . . . and that it will occupy a proud place in your library. We would like to keep you informed about other publications from Schiffer Publishing Ltd.

TITLE OF BOOK: _____

☐ hardcover
☐ paperback

☐ Bought at: _____
☐ Received as gift

COMMENTS: _____

_____

☐ *Please send me a free Schiffer Arts, Antiques & Collectibles catalog.*

☐ *Please send me a free Schiffer Woodcarving, Woodworking & Crafts catalog*

☐ *Please send me a free Schiffer Military /Aviation History catalog*

☐ *Please send me a free Whitford Press Mind, Body & Spirit and Donning Pictorials & Cookbooks catalog.*

Name _____

Address _____

City _____ State _____ Zip _____

SCHIFFER BOOKS ARE CURRENTLY AVAILABLE FROM YOUR BOOKSELLER

Aunt Jemima in her finer attire, with bunnies on her apron —surely "After the Receipt was Sold." Marked "Aunt Jemima" on back of collar, 1929. *Author's Collection.*

Cut and unsewn dolls can easily be displayed in frames, between two pieces of UV protective glass, where they are less likely to be damaged by fading and normal household air pollutants.

Left and Right:
Uncle Mose with
a bright orange
shirt and blue
pants; one hand
is in his pocket,
the other holds a
pipe. Marked
"Uncle Mose/Aunt
Jemima's/
Husband," 1924.
Author's Collec-
tion.

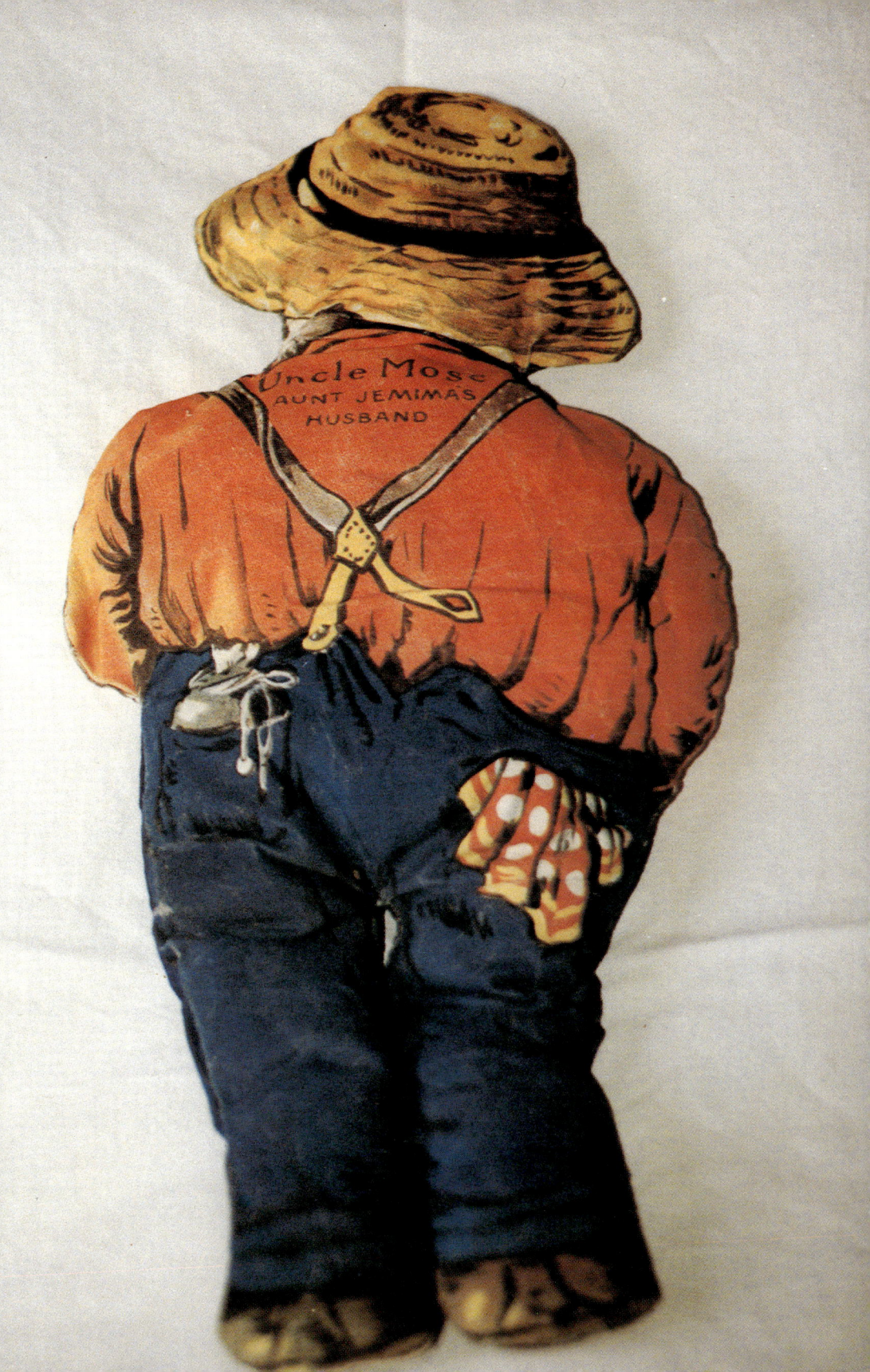

Uncle Mose with a pipe in his right pocket, holding his left suspender. Marked "Uncle Mose" across shoulders on the back of his shirt, 1924. *Photograph courtesy of Myla Perkins.*

Opposite page:
Uncle Mose in more subdued colors, with a red shirt, gray and black striped pants, a pipe in his left pocket and his arms crossed. The soles of his feet are sewn separately. Marked "Uncle Mose" on back, 1924. *Photograph courtesy of Myla Perkins.*

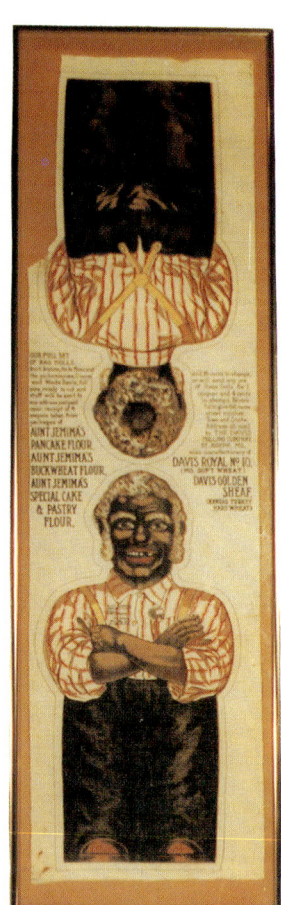

Opposite page:
Wade, holding a straw hat and with a pointing index finger on his right hand, from Aunt Jemima Mills Co. Marked "Wade Davis" on top of pants in back, 1924. *Photograph courtesy of Myla Perkins.*

Uncle Mose smiling, with teeth missing, in a red and white striped shirt, arms folded, with a spoon in his left hand and a pipe in his left pocket. Marked on front right chest, "Aunt Jemima's Pancake Flour/Pickanniny Doll/Uncle Moses/The Davis Milling Co., St. Joseph, MO," 1905. *Photograph coutesy of Myla Perkins.*

Uncle Mose, finely attired with a top hat and long jacket—"After the Receipt was Sold." Marked "Uncle Mose" on back of collar, 1929. *Photograph courtesy of Myla Perkins.*

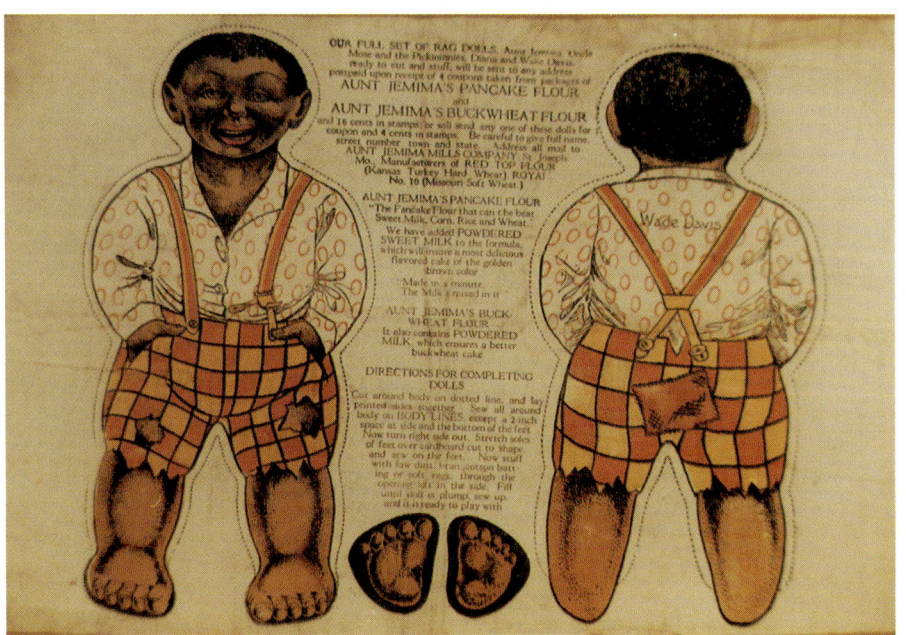

Wade with his hands in his pockets, white shirt with red circles, and checkered short pants. Soles of his feet are separate. Marked "Wade Davis" on back of shirt, 1924. *Photograph courtesy of Myla Perkins.*

Diana wearing a yellow polka-dot dress, with three tufts of hair and her right index finger to her lip. She has plump arms and very disproportionately thin legs, which are together. Marked "Aunt Jemima's/Pancake Flour/Pickaninny Doll/Diana/The Davis Milling Company/St. Joseph, MO," 1905. *Photograph courtesy of Myla Perkins.*

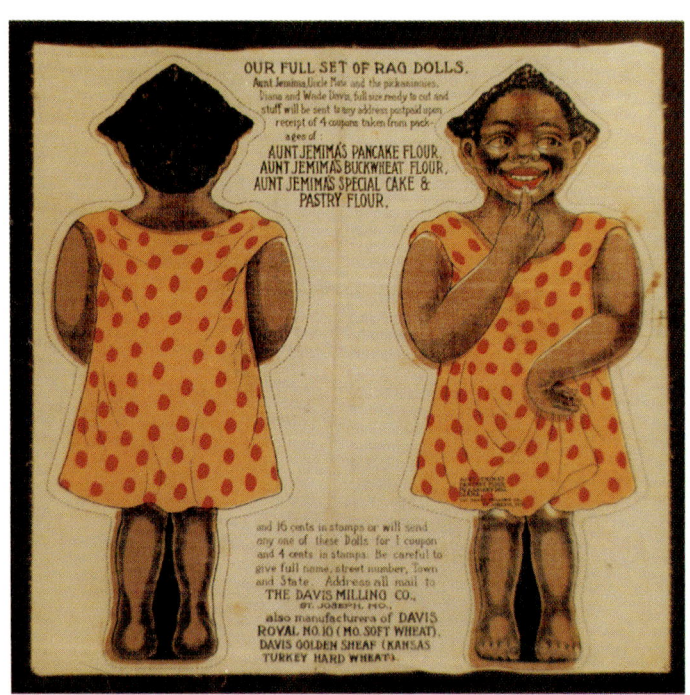

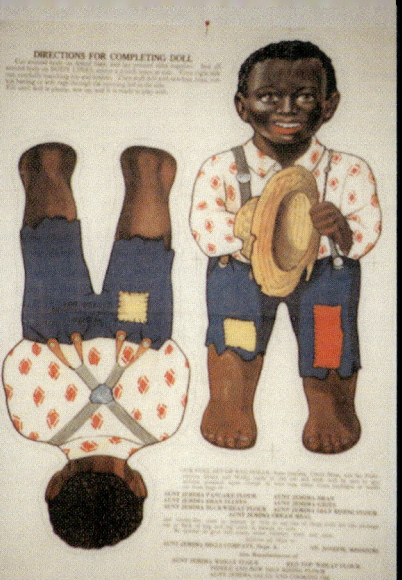

Wade wearing dark blue pants and holding a straw hat; does not have separate soles. Marked "Wade/Aunt Jemima's/Little Boy," 1924. *Author's Collection.*

Diana smiling in a blue, daisy-print dress, holding a kitten. Marked "Diana/ Aunt Jemima's/Little Girl" on back of her dress, 1924. *Author's Collection.*

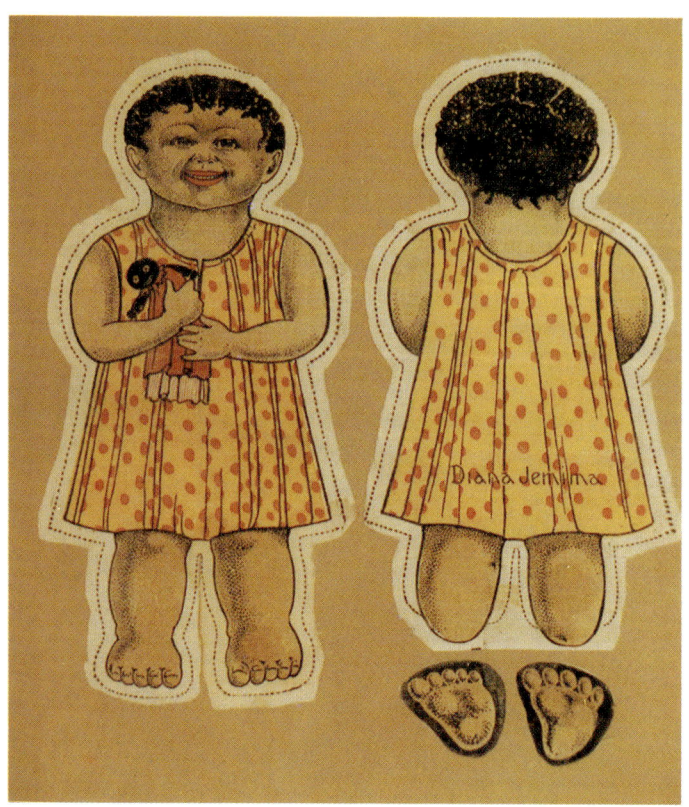

Diana wearing a polka-dot dress and holding a black doll. The soles of her feet are sewn separately. Marked at bottom of dress on the back, "Diana Jemima," 1924. *Photograph courtesy of Myla Perkins.*

Diana and Wade prototypes, or possibly transfers which were printed on the back of early thick, brittle paper which resembles wallpaper, late 1940s. *Author's Collection.* $50-$75/pair.

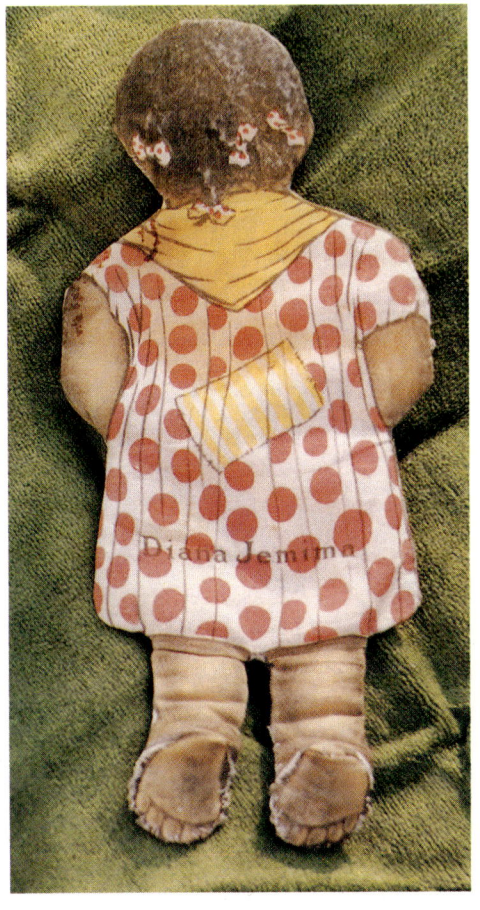

A smiling Diana doll, wearing a scarf and a white dress with red polka dots and yellow patches, holding a kitten. The soles of her feet are sewn separately. Marked in black, "Diana Jemima," at the bottom of the back of the dress. *Author's Collection.*

Pancake Mix box with oilcloth doll offer. Diana and Wade dolls were FREE when a customer ordered Aunt Jemima and Uncle Mose dolls for 50 cents. *Burkett Collection.* $25-$50.

A later oilcloth Rag Doll Family coupon offer shows Aunt Jemima and Uncle Mose at 25 cents each and Diana and Wade at 25 cents for both. Inflation! *Burkett Collection.* $15-$20.

A Sunday funny paper advertisement showing oilcloth Rag Doll Family. "Offer closes January 15, 1950." *Author's Collection.* $15-$20.

Full family of oilcloth dolls, unstuffed. *Author's Collection.* $300-$350/set, depending on condition.

CHAPTER 5

F & F Mold & Die Co. Premiums

Opposite page:
Red plastic black face cookie jar by F & F. *Author's Collection.* $400-$450.

Blue plastic black face cookie jar by F & F Mold & Die Co, ca. 1951. A recent auction price for a green jar in mint condition was $2000. *Burkett Collection.*

Order blank for a cookie jar for $1.00. On the reverse is an order form for Aunt Jemima Mother Goose Candle Holders for cakes.

Brown face cookie jar with original mailing box. *Burkett Collection.* Box alone $75-$100.

The side view of Aunt Jemima Ready-Mix Buckwheat, Wheat and Corn Flour Pancake box, with a cookie jar offer.

Aunt Jemima six-piece spice set in red plastic, manufactured by F & F Mold & Die
Co., Dayton, OH, ca. 1949. *Author's Collection.* $300-$350.

"WAKE UP to Aunt Jemima Pancakes" advertisement from *The Saturday Evening
Post*, September 16, 1950. $20-$25.

Coupon for ordering three Aunt Jemima spices, ca. 1951. $20-$25.

Grocery store poster promoting three pieces of a six-piece spice set for 50 cents, ca. 1950. 16-1/2" x 22". *Burkett Collection.* $200-$250

Copper rack for the six-piece spice set with Mississippi Queen River Boat and original mailing box. This was not manufactured by F & F Mold & Die Company. *Author's Collection.* $450-$500 with mailing box; $400-$450 rack only.

This Aunt Jemima plastic sugar with lid and Uncle Mose creamer were boxtop coupon offers manufactured by F & F Mold & Die Co., Dayton, OH. They were offered in yellow, blue, and green. The yellow, most common, were issued in 1949, and the blue and green in 1950. The blue and the green were prototypes and therefore are now quite rare. *Burkett and Author's Collection.* Yellow $125/set; green and blue $200-250/set.

Aunt Jemima Creamer and Sugar set order coupon, ca. 1952. *Burkett Collection.* $20-$25.

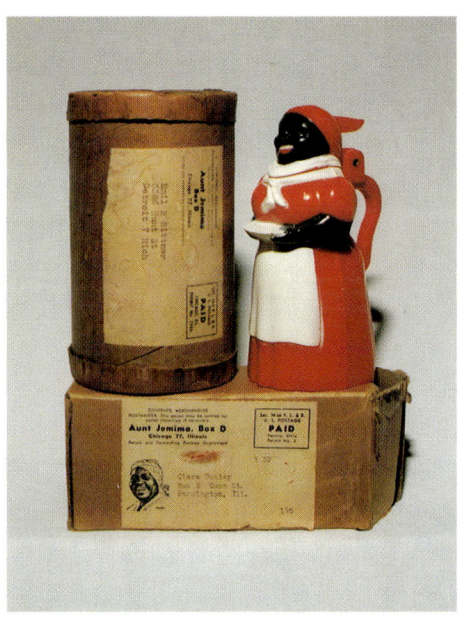

Aunt Jemima Syrup Pitcher were mailed in two different types of boxes. *Burkett Collection.* Mailing boxes with label, $25-$35.

A Sunday funny paper advertisement for Syrup Pitcher "Start the Day this Gay Way," ca. 1949. *Author's Collection.* $15-$20.

Coupons for Aunt Jemima Syrup Pitcher, ca. 1948. *Burkett Collection.* $20-$25.

Opposite page:
Aunt Jemima Syrup Pitcher, manufactured by F & F Mold & Die Co., Dayton, OH, ca. 1949. Originally a boxtop coupon offer, it was also attached to boxes of pancake mix. 5-1/2". *Author's Collection.* $55-$65.

Plastic salt and pepper shakers manufactured by F & F Mold & Die Co., Dayton, OH, ca. 1948. The smaller size was advertised as "Dainty" table size. 3-1/2" and 5". *Author's Collection.* Small $40-$45; large $50-$55.

Above and Right: Coupons for ordering salt and pepper shakers, ca. 1948. *Burkett Collection.* $20-$25.

This "Cake 'n' Ice Cream Spoon" was a give-away inside specially marked packages of Aunt Jemima Cake Mixes. There were twelve different Mother Goose characters in all. Manufactured by F & F Mold & Die Co., Dayton, OH, early 1950s. *Author's Collection.* Spoons $50-$75/set; magazine ad $20-$25.

Mother Goose Candle Holders manufactured by F & F Mold & Die Co., Dayton, OH. These were a boxtop premium—"For each set of 12 I enclose 25 cents and the top from one package of Aunt Jemima Silver Cake or Devil's Food," ca. 1951. Reverse of coupon is an order blank for red plastic cookie jar. *Author's Collection.* $40-$45/set; with mailing box $75-$100; coupon for ordering $15-$20.

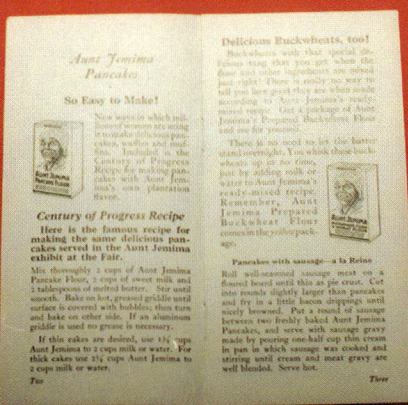

"America's Most Famous Recipe" with a 'live' Aunt Jemima serving pancakes on a square paper plate. Front cover of booklet shows a black and white picture of Aunt Jemima in a log cabin kitchen with pancake mix boxes in the background, ca. 1933. Inside is the 'Century of Progress' recipe from the World's Fair. 6" x 3". *Burkett Collection.* $25-$50.

Opposite page:
"A Recipe No Other Mammy Could Equal," includes the story of Aunt Jemima's rise to fame and some recipes, copyright 1928. 6" x 3". *Author's Collection.* $50-$75.

AUNT JEMIMA

AMERICA'S
MOST FAMOUS
RECIPE

THE QUAKER OATS COMPANY

CHICAGO, ILLINOIS

Recipe booklets, "Adventures in Corn Meal Cookery!" "Aunt Jemima's New Temptilatin' Menus and Recipes," and "Aunt Jemima's New Temptilatin' Recipes," ca. 1950s. 5-1/2" x 4". *Burkett and Author's Collections.* $25-$35 each.

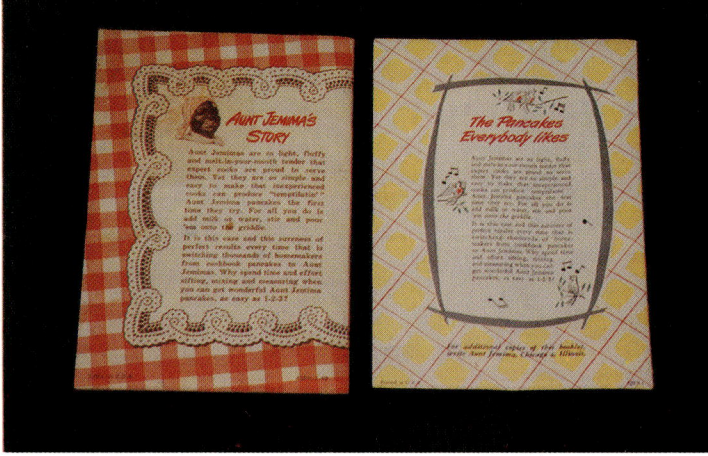

Aunt Jemima's LENTEN Meals tear-off coupon from grocers' shelves. Imagine waffles with creamed tuna! *Burkett Collection.* $15-$20.

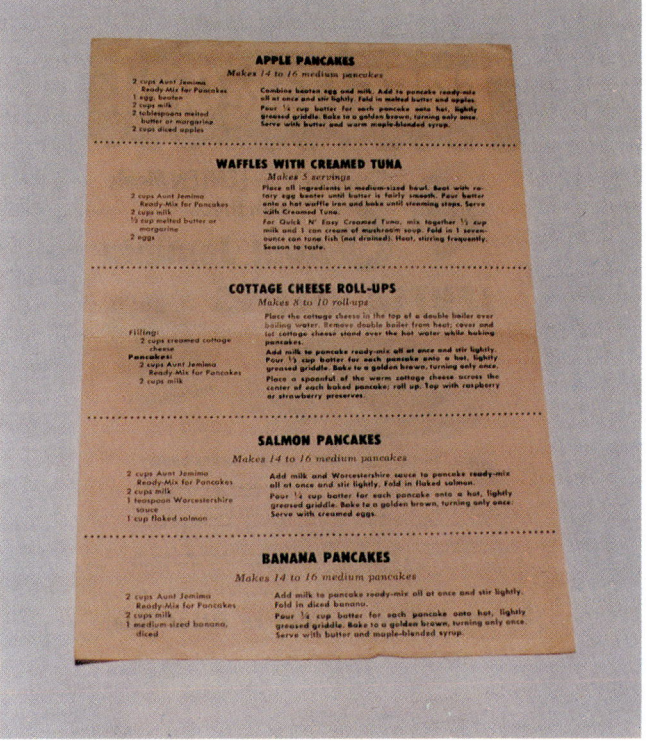

Opposite page:
"Aunt Jemima's Album of Secret Recipes" booklet, in color. $10-$15.

"12 Magic Menus from Aunt Jemima's Album of Secret Recipes" booklet. Printed in black and white. $10-$15.

"Aunt Jemima's Magical Recipes" booklet is found with different copyright dates; the earliest is 1952, and another is 1954. $25-$30.

Aunt Jemima's
Album of Secret Recipes

"I'se in town, Honey!"
America's most famous recipe
New ways in which millions of women are using it to make delicious pancakes, waffles and muffins
"I'se in town, Honey!"
America's most famous recipe
New ways in which millions of women are using it to make delicious pancakes, waffles and muffins

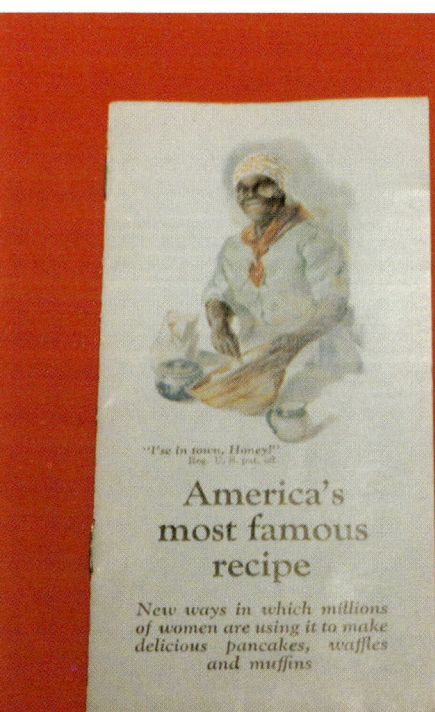

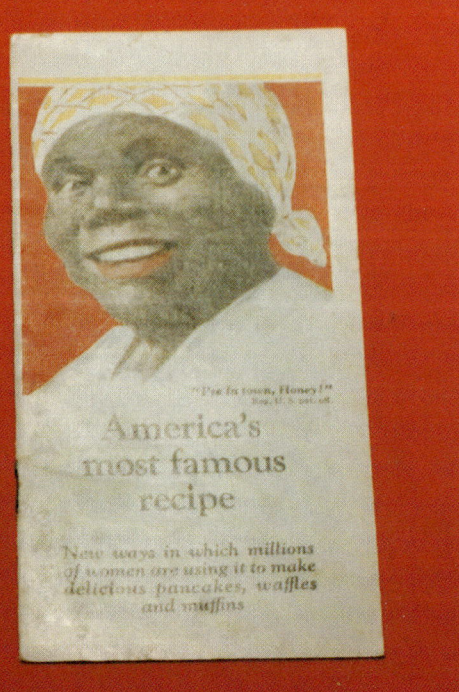

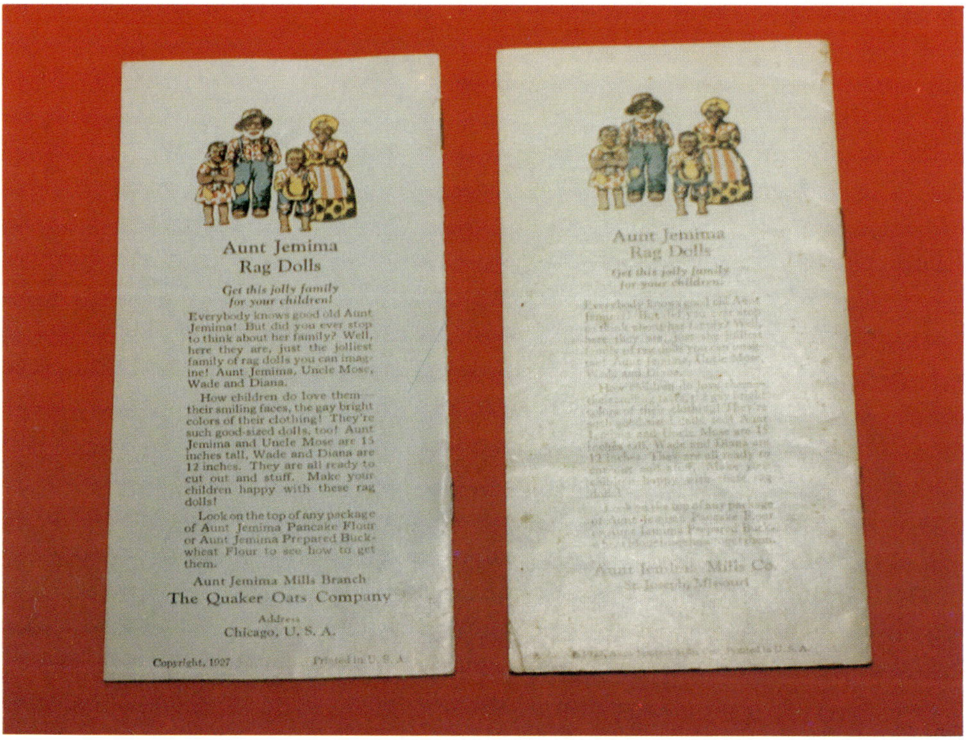
Aunt Jemima
Rag Dolls
Get this jolly family for your children!
Everybody knows good old Aunt Jemima! But did you ever stop to think about her family? Well, here they are, just the jolliest family of rag dolls you can imagine! Aunt Jemima, Uncle Mose, Wade and Diana.
How children do love them—their smiling faces, the gay bright colors of their clothing! They're such good-sized dolls, too! Aunt Jemima and Uncle Mose are 15 inches tall, Wade and Diana are 12 inches. They are all ready to cut out and stuff. Make your children happy with these rag dolls!
Look on the top of any package of Aunt Jemima Pancake Flour or Aunt Jemima Prepared Buckwheat Flour to see how to get them.
Aunt Jemima Mills Branch
The Quaker Oats Company
Address
Chicago, U. S. A.
Copyright, 1927
Aunt Jemima
Rag Dolls
Get this jolly family for your children!

Opposite page:
Two versions of the recipe booklet "America's Most Famous Recipe," with the folklore tale inside and a few recipes, ca. 1927 (left), and ca. 1925 (right). *Burkett Collection.* $50-$75 each.

"Pancakes Unlimited" booklet with several pancake recipes and scenes using the blue pancake shaker premium, ca. 1958. *Author's Collection.* $15-25.

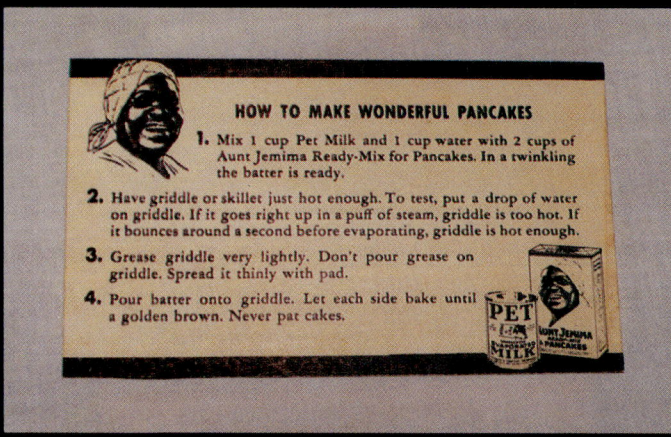

HOW TO MAKE WONDERFUL PANCAKES

1. Mix 1 cup Pet Milk and 1 cup water with 2 cups of Aunt Jemima Ready-Mix for Pancakes. In a twinkling the batter is ready.

2. Have griddle or skillet just hot enough. To test, put a drop of water on griddle. If it goes right up in a puff of steam, griddle is too hot. If it bounces around a second before evaporating, griddle is hot enough.

3. Grease griddle very lightly. Don't pour grease on griddle. Spread it thinly with pad.

4. Pour batter onto griddle. Let each side bake until a golden brown. Never pat cakes.

A 3" x 5" card on how to make wonderful pancakes using PET MILK. *Burkett Collection.* $10-$15.

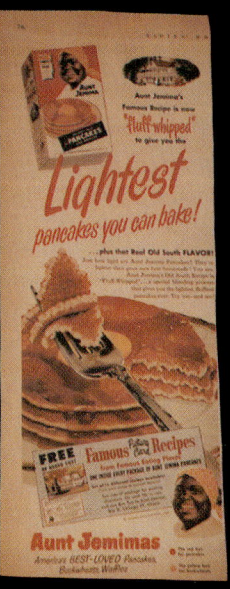

Aunt Jemima Recipe Picture Cards with famous eating places featured on front and recipe on reverse. Artist's name noted on front of each picture, ca. 1952. 3" x 5". *Author's Collection.* $125-$150/set.

Magazine advertisement for Famous Recipe Picture Cards—"one inside every package of Aunt Jemima Pancakes," or send 10 cents in coin and one boxtop. Marked "Quaker Oats Company," ca. 1952. *Author's Collection.* $15-$20.

CHAPTER 7
Aunt Jemima's Kitchen

Picture postcards of two Aunt Jemima's Kitchen Restaurants, mid-1960s. The restaurant in the top picture is located in Grand Rapids, MI. Other location unknown. *Burkett Collection.* $35-$50.

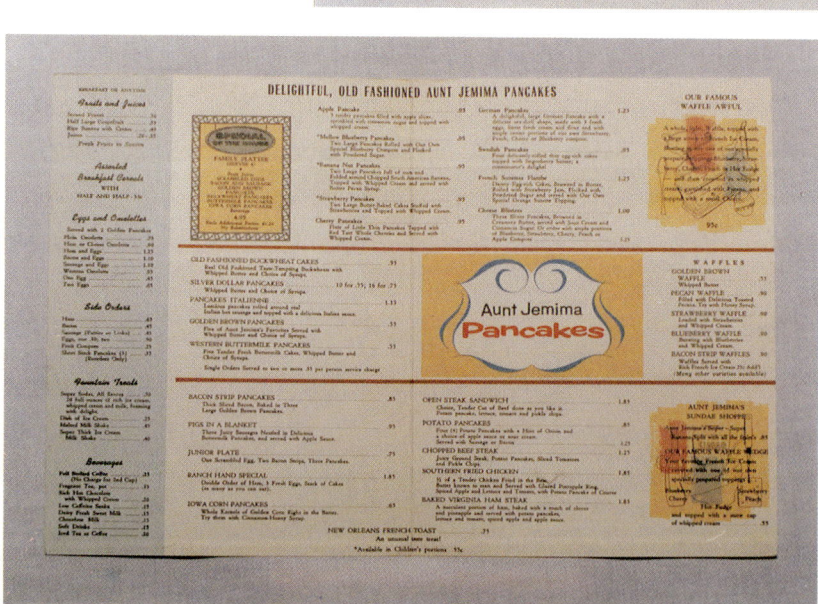

Aunt Jemima's Kitchen Menu lists a full line of dining, from pancakes to southern fried chicken to sundaes. 6-3/4" x 5". *Burkett Collection.* $35-$50.

Menu from Aunt Jemima's Kitchen at Frontierland in Disneyland. *Burkett Collection.* $15-$25.

Several examples of paper placemats used at various promotional functions and at Aunt Jemima Kitchen Restaurants. Note: Page 75 bottom left is in Spanish. Dates on these vary. Each measures 10" x 14". *Burkett Collection.* $10-$20 each.

The Story of AUNT JEMIMA
The Story of Aunt Jemima's Kitchen
"It's always a treat to eat at Aunt Jemima's Kitchen"

Aunt Jemima
AS DISTINCTIVE AS THE OLD SOUTH

Paper napkins from the restaurant and from promotional events. 7" square. *Burkett Collection.* $15-$25 each.

"The Legend of Aunt Jemima" napkin opened to show her folklore tale.

Top: Plate from the 1933 World's Fair. 8" square. *Burkett Collection.* $50-$75.
Bottom two: Round paper plates with the Aunt Jemima logo. 9" diam. (left) $25-$50,
(right) $15-$25/each.

Paper cup, early 1960s. 6 oz., 3-3/4" high. *Burkett Collection.* $20-$40 each.

Aunt Jemima's Restaurant Placecard.
Author's Collection. $20-$25.

Matchbook from the Disneyland
restaurant. *Burkett Collection.* $20-$25.

Aunt Jemima Restaurant Dinnerware included 7" and 10" plates, three different size bowls—a grits or fruit bowl (the smallest), a soup bowl, and a cereal bowl—and a cup and saucer. Plates $75-$100 each, bowls $50-$100 each, cup and saucer $150-$175/set.

CHAPTER 8
Children's Toys

Junior Chef Pastry Mix Set with Aunt Jemima Ready-Mixes, and Junior Chef Pastry and Pancake Mix Set. Both sets included cookbook, Aunt Jemima cake mixes, baking pans, cookie cutter, baking sheet, and rolling pin. These sets were offered in the 1955 Sears catalog. *Burkett Collection.* $100-$150 each.

The Junior Chef Pancake Making Kit included measuring spoons, measuring cups, wire whisk, batter squeezer tube, heat resistant pancake turner, serving plate, and syrup pitcher, ca. 1988. *Author's Collection.* $75-$100.

Junior Size Aunt Jemima Mix packages from Pastry Mix sets. *Author's Collection.* $20-$25 each.

Opposite page: "BREAKFAST BEAR" offer appeared only on Aunt Jemima Frozen Waffles boxes. 13". *Burkett Collection.* $50-$75 if complete.

Baking Mix Set, featuring Aunt Jemima and Flako products, included pie plate and cake pan, muffin pan, and Junior mixes, early 1960s. *Burkett Collection.* $100-$125.

Balsa Wood Airplane with original mailing box. Wings have Aunt Jemima logo and "Wright-Dayton/Made for/Aunt Jemima," ca. 1930s. Wing span is 14". *Burkett Collection.* $N/A.

"HOORAY! It's Aunt Jemima Day" coloring booklet, ca. 1953. Notice the yellow pancake shaker on the cover. Inside shows children making pancakes with the 'pancake mold' and serving them to Mom and Dad. *Burkett Collection.* $50-$75.

Aunt Jemima Pancake Jamboree Banner. Bright yellow with red lettering and the full face logo on both sides. 30" long x 17" high. *Author's Collection.* $150-$200.

The Aunt Jemima Pancake Griddle featured a water tank under the embossed Aunt Jemima bust (for filling a measuring cup with the exact amount of water for the mix), a beater (missing from this piece), and six rings on the griddle to pour batter into. Used for grocery store and fair demonstrations. Has a Quaker Oats metal tag and serial number. 26" high x 24-1/2" wide. *Burkett Collection.* $N/A.

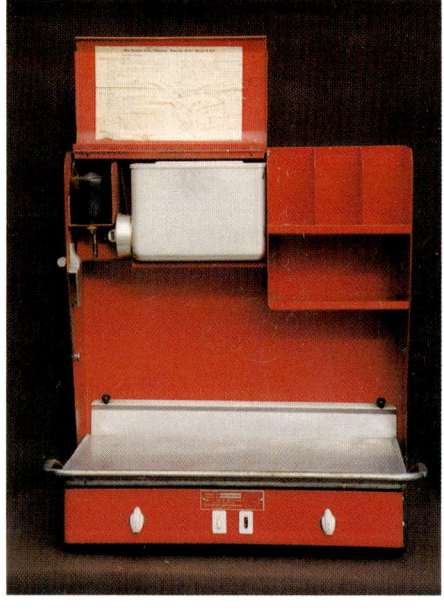

Newspaper advertisement for Lutterman's Super Market in Mt. Vernon, IN promoting a personal appearance of Aunt Jemima, ca. 1957. *Burkett Collection.* $N/A.

Aunt Jemima Pancake Jamboree Apron, ca. 1970s. 36" x 22". *Burkett Collection.* $50-$75.

Aunt Jemima Paper Hats were give-aways at different promotional events, ca. 1953 (top), ca. 1958 (middle), ca. 1966 (bottom). All are 3-1/2" x 11". *Burkett Collection.* $25-$50 each.

Aunt Jemima Personal Appearance
Banner for the Kiwanis Pancake
Festival, dated 1951. 34" x 57". *Burkett
Collection.* $350-$450.

Banner announcing Aunt Jemima
Pancake Jamboree event to be held
Tuesday, September 26, ca. 1970. 34" x
55". *Burkett Collection.* $100-$200.

Metal tab pins, Aunt Jemima Pancake Club and Aunt Jemima Breakfast Club, early 1950s. 2-1/4" each.

Aunt Jemima Campaign Chairman pin give-away at store and fair demonstrations, early 1950s. 1" diam.

Aunt Jemima Breakfast Club metal pin, also a give-away at demonstrations. 4" diam.

Note: These have been reproduced in recent years. Original pinbacks are marked "Green Duck Co." or "Adcraft Co.," both of Chicago, IL. Green Duck Co. also produced a 3" pinback with the same logo.

CHAPTER 10
Miscellaneous Collectibles

Aunt Jemima's Magic Pantry store display. 20-1/2" x 17" x 13-1/2". *Burkett Collection.* $400-$500.

Aunt Jemima's Magic Shelf grocery display, ca. 1936. 21" x 17". *Burkett Collection.* $350-$450.

Square, easel-back sign, "Folk Comes Miles to Eat Our Pancakes" with Aunt Jemima logo in lower right corner, and the caption "They're Genuine Aunt Jemima's," ca. 1949. 12" x 17". *Burkett Collection.* $50-$75.

Breakfast Bonus grocer's promotion sign, ca. unknown. 10-1/4" x 17". *Burkett Collection.* $50-$75.

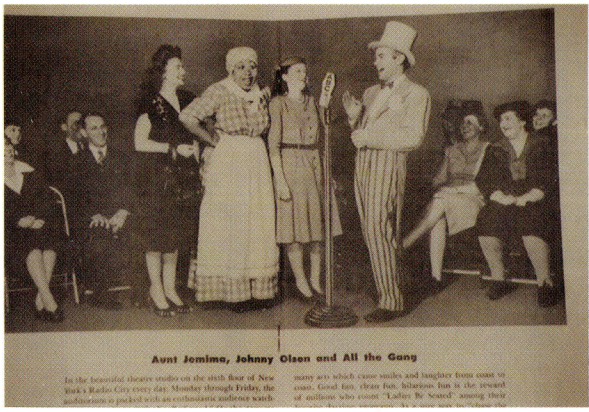

Aunt Jemima, Johnny Olsen and All the Gang

Souvenir program from "Ladies Be Seated" radio show, starring Aunt Jemima, "The Pancake Queen," Jonnny Olsen, and All the Gang, ca. unknown. Could this be the same Johnny Olsen from "The Price is Right"? 5-1/4" x 8-1/2". *Author's Collection.* $50-$75.

Aunt Jemima Pancakes grocery store poster, mid to late 1950s. 19" x 26". *Burkett Collection.* $75-$100.

Premium Redemption Gift Catalog for premiums given with Mother's Oats and Aunt Jemima Pancake Flour. Inside you can order the 1924 rag doll family set for 30 coupons, or for 2 coupons and 25 cents. Dated October, 1926. *Author's Collection.* $50-$75.

"Simple Simon's School House" child's booklet printed in conjunction with a radio program featuring Phil "The Colonel" Cook, sponsored by by Quaker Oats Co. of Chicago, ca. unknown. *Author's Collection*. $35-50.

"Well Youngsters...here's the Book I've been talking about these last few weeks on the Radio...Just a few of Simple Simon's Crazy little School House Rhymes..."

"Now I tink dats enough Foolishness in de school house... and de Poopils will all go home and for Breakfast Tomorrow Morning...de Mother Goose is Having...Aunt Jemima' Pancakes with Syrup and melted Butter...dats what I call Appateasin...a appa...a...Dots a delicous Breakfast.

"...don't put Popcorn in the Pancake Batter and expect the pancakes to turn by themselves...use Aunt Jemima....instead."

Blow-molded plastic referred to as "soft" cookie jar. *Burkett Collection*. $200-$225.

Cardboard die-cut "I'se in Town Honey," Aunt Jemima's Pancake Flour sign, early 1900s. 14" x 7-1/2". *Burkett Collection.* $N/A.

"I'SE IN TOWN HONEY"

AUNT JEMIMA'S PANCAKE FLOUR

Advertising blotter card with "I'se in Town Honey" logo, ca. 1920s. *Burkett Collection.* $75-$100.

Paper grocer's window sign. *Burkett Collection.* $10-$15.

Cardboard poster featuring Aunt Jemima and two disc jockeys from WNJR Newark, "America's Greatest Rhythm and Blues Station." 22" x 14". *Burkett Collection.* $25-$35.

Tear-off coupon from the grocer's with buttermilk pancake recipe on the front and a coupon for free Aunt Jemima Pancake Mix on the reverse—send bottle cap or portion of a buttermilk carton which says "Buttermilk" and a box top from Aunt Jemima Pancake Mix. *Burkett Collection.* $10-$15.

Glass Slide used in theaters by local merchants to advertise, ca. 1930s. This merchant is Bailey & Son, who bought rights to advertise in the space under Aunt Jemima. Made by the Kansas City Slide Mfg. Co. 3-1/2" x 4". *Burkett Collection.* $N/A.

(Above) Trolley car ad, ca. 1919. 21" x 11". *Burkett Collection.* $250-$300.
(Below) Trolley car ads, ca. 1940s. 21" x 11". *Burkett Collection.* $200-$250 each.

Round metal pancake mold with four animal cutouts. Marked "Aunt Jemima" on reverse. 8-1/2" diam. *Author's Collection.* $155-175.

Cigarette lighters: (left) Aunt Jemima Frozen Foods, mfg. by Cobid; (right) Aunt Jemima with full face logo in red enamel, mfg. by Dundee. *Burkett Collection.* $50-$75 each.

Lite Syrup Introduction/ Presentation Set, possibly used by distributors, mid-1980s. *Burkett Collection.* $20-$25.

Yellow plastic pancake shaker with full face logo on top and "Perfect Pancakes in 10 Shakes," ca. 1948. 8-3/4" high. *Author's Collection.* $50-$60.

Aunt Jemima Deluxe Easy Pour box with intructions on reverse showing yellow pancake shaker. *Author's Collection.* $20-$25.

Blue plastic Pancake Shaker was a boxtop premium to promote the new "Deluxe Buttermilk Pancake Mix." Offer required three boxtops, one each from the Buttermilk, Buckwheat, and Plain Pancake Mixes. Offer expired December 31, 1957. *Author's Collection.* Shaker $50-$65, advertisement $10-$15.

Glass ball jar with screw-on metal lid. Jar has marks for 2-1/2 cups, which is total for all ingredients. Underneath the lid are directions for the mix, ca. 1950s. *Author's Collection.* $75-$100.

Aluminum pancake griddle for stovetop, complete with flyer. No markings on the pan itself, ca. 1952. 12-1/2" diam. *Burkett Collection.* $50-$75 with flyer.

Aunt Jemima Mills letter opener with knife, ca. 1914-1925. 8-1/2" long. *Burkett Collection.* $N/A.

Aunt Jemima Batter Spoon was a boxtop premium plus 10 cents, as advertised in the Sunday funny paper section. Ad offer closed February 22, 1948. Spoon is unmarked. *Author's Collection.* $N/A.

Aunt Jemima Pancake Flour green coupon, interchangable with coupons packed in Mother's Oats and Mother's Flour. "Redeem 2 coupons plus 20 cents to receive Community Tudor Plate cereal spoon, or twenty-three coupons alone. If preferred 1/4 cent apiece when sent in lots of not less than 40." Not redeemable after October 1, 1929. *Burkett Collection.* $15-$20.

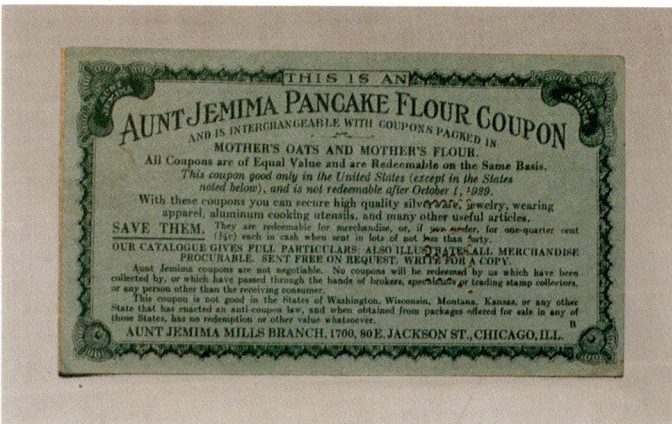

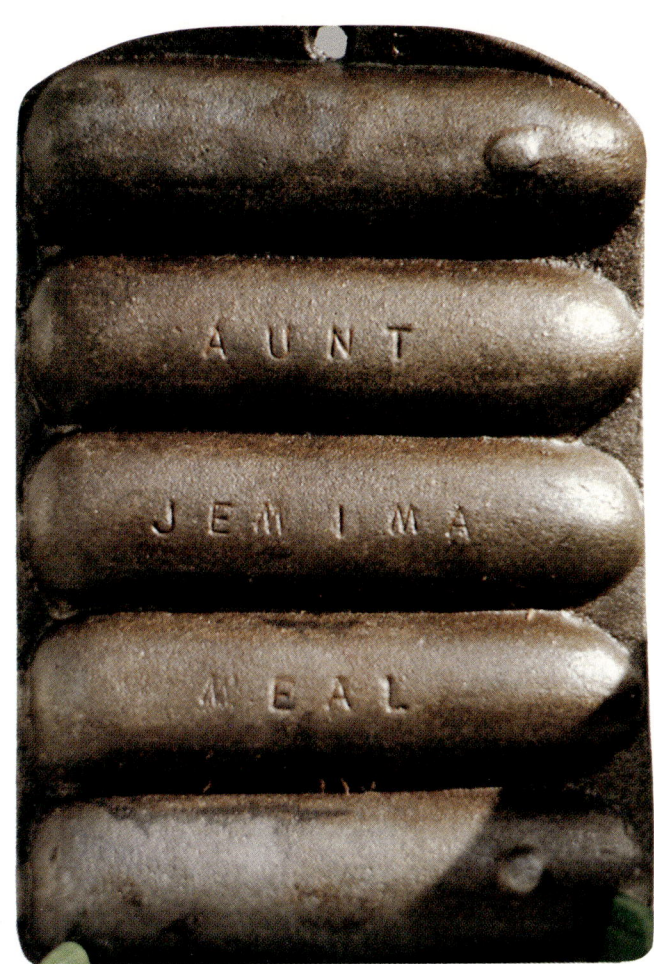

Corn muffin pan marked "Aunt Jemima Meal" on back. Unknown to author or contributors how this premium was obtained. *Author's Collection.* $50-$75.

 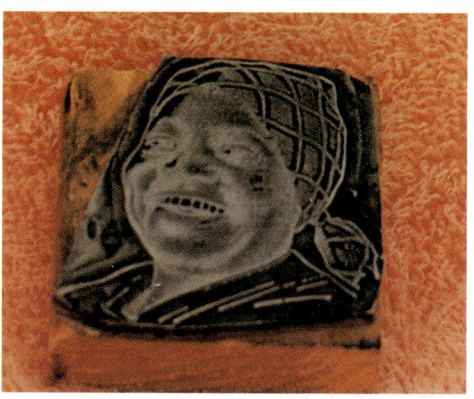

Printer's cuts of full face logo and product boxes for magazine and newspaper advertising. *Author's Collection.* (right) 2" x 2", $75-$100; (left) 1-1/4" x 3/4", $50-$75.

Round metal ashtray reads around the edge: "Aunt Jemima's Flour, Aunt Jemima Meal in Sacks, Aunt Jemima Grits and Meal in Packages, Red Top Flour." Center reads: "Aunt Jemima Mills Branch, The Quaker Oats Co., St. Joseph, MO." *Burkett Collection.* $N/A.

Aunt Jemima Pancake die-cast metal commemorative bank by Ertl, ca. 1992. Dates 1928-1992 apply to the Quaker Oats plant in Cedar Rapids, Iowa, not the plant in St. Joseph, MO which had acquired the Aunt Jemima product in 1926. *Burkett Collection.* $75.

Aunt Jemima Pancakes Apron, ca. unknown. *Burkett Collection.* $25-$50.

Canvas tote bag, apron and golf shirt, with pre-1968 full face logo, were available to employees of the Quaker Oats Co. at the employee store, ca. 1992. *Author's Collection.* $20-$25 each.

Work gloves marked "AUNT JEMIMA The Breakfast Expert" were given to employees of Quaker Oats Co. The pot holder was attached to bottles of pancake syrup. Measuring spoons were mail-in premiums. The spatula is the most recent premium, from 1993, and was attached to syrup bottles. *Author's Collection.*

The plastic cannister by Rubbermaid and plastic measuring cups came with bottles of pancake syrup.

A ceramic syrup pitcher boxtop premium from 1988 required 75 cents for postage and handling and 4 UPC/purchase seals. $20-$25.

A very early Aunt Jemima Pancake Syrup/Sugar Cream Recipe Booklet. Products manufactured by Rigney & Co., Brooklyn, NY, under the Aunt Jemima trademark logo and name. Pre-1916. *Author's Collection.* $75-100.

Front and back of a tin cannister for pancake flour, a mail-in promotion from 1983. 5-1/4" dima. x 6-1/4" high. *Burkett Collection.* $25-$50.

BIBLIOGRAPHY

Marquette, Arthur F. *Brands, Trademarks and Goodwill: The Story of the Quaker Oat Company,* undated, McGraw-Hill Book Company.

Robinson, Joleen Ashman and Kay F. Sellers. *Advertising Dolls, Identification & Value Guide,* 1980 Collector Books. Paducah, KY.

Wright, Purd. "Aunt Jemima, The Most Famous Colored Woman in the World," written for R. T. Davis Mills, 1895, St. Joseph, MO.

Aunt Jemima Frozen Waffle Clock, distributor's premium, ca. 1989-1990. 4" x 7" high. *Burkett Collection.* $50-$75.

DAVIS MILLS IN 1845.

DAVIS MILLS IN 1895.

Compliments of
R.T. DAVIS MILL CO.
ST. JOSEPH, MO.